DATE			

Library of Congress Cataloging-in-Publication Data

Watts, Barrie.
    Tomato / Barrie Watts.
        p.   cm.—(Stopwatch)
    "First published by A & C Black...Adapted and published in the
United States in 1989 by Silver Burdett Press"—T.p. verso.
    Summary: Follows, in text and illustrations, the development of a
tomato from seedling to full maturity.
    1. Tomatoes–Juvenile literature.   [1. Tomatoes.]   I. Title.
II. Series: Stopwatch books.
    58349.W36   1989
    635.642-dc20   89-38982
        ISBN 0-382-24010-3   CIP
        ISBN 0-382-24008-1 (lib. bdg.)   AC

First published by A & C Black (Publishers) Limited
35 Bedford Row, London WC1R 4JH

© 1989 Barrie Watts

Adapted and published in the United States in 1990
by Silver Burdett Press, Englewood Cliffs, New Jersey
U.S. project editor: Nancy Furstinger

Acknowledgments
The artwork is by Helen Senior.
The publishers would like to thank Michael Chinery for his help and advice;
also the Institute of Horticultural Research, Littlehampton.

Printed in Belgium by Proost International Book Production

Tomato

Barrie Watts

Silver Burdett Press • Englewood Cliffs, New Jersey

Here are some tomatoes.

Tomatoes can be big or small, red or orange. Some of them are even yellow.

Tomatoes taste good cooked or raw. Do you like tomato sauce on your spaghetti?

Look at the photograph below. These tomato plants are growing in a greenhouse. Some people grow tomato plants in their gardens.

This book will tell you how tomatoes grow.

A tomato has seeds.

This tomato has been cut in half. Inside, it is soft and juicy. Can you see the seeds? Each tomato seed is as small as a pinhead.

This seed is dry. It is covered by little hairs.

The seed starts to grow.

You can plant the seed in your garden. When spring comes, it will start to grow. Look at the drawing.

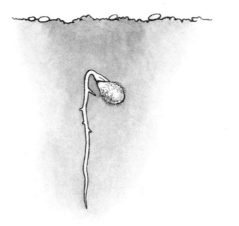

A root is growing. It has pushed down out of the seed.

After a week, a little shoot starts to grow. Look at the photograph. The shoot has grown up above the ground.

Can you see the green leaves near the top of the shoot? There are two leaves pressed together. At the top of the leaves is the seed coat.

The first leaves open.

The tomato plant is eight days old. The seed coat drops off and the two leaves open. Can you see the new shoot between the leaves? The stem and leaves are hairy.

The leaves spread out to catch the light.
The plant needs sunlight, air, and water to grow.

Look at the drawing. The plant has grown more roots.

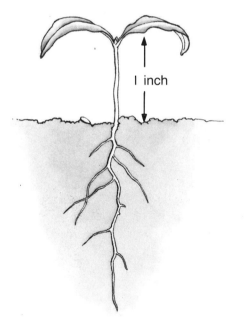

I inch

The roots take in water and food from the soil.

The plant grows more leaves.

The tomato plant has been growing for two weeks. It has grown branches with big leaves. You can still see the first two leaves at the bottom of the plant. But soon they will drop off.

Here are some big tomato leaves.

These big leaves are a different shape from the first two leaves. The big leaves have lots of tiny tubes called veins. The veins carry food to all parts of the plant.

The plant grows flower buds.

When the tomato plant is five weeks old it grows flower buds. The buds grow on small branches coming out from the main stem.

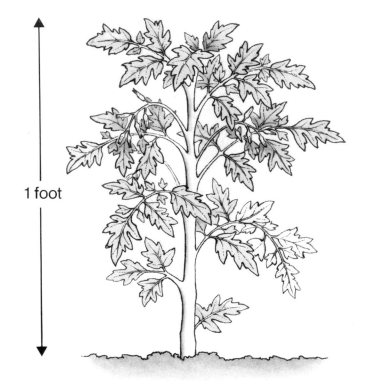

1 foot

Each bud is wrapped in green sepals (tiny leaves). The buds are covered with hairs to stop insects from eating them.

The flower buds open.

Soon the sepals split apart and bend back. The flower buds begin to open one by one. Then the tomato flowers bloom.

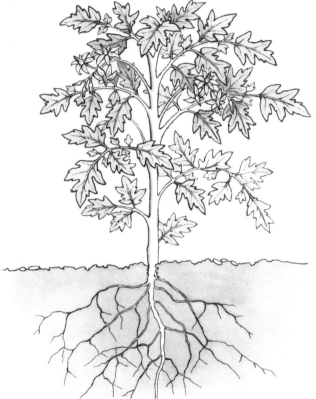

The flower petals curl back and the middle part of each flower sticks straight out. In this part is a yellow dust called pollen.

Pollen falls inside a flower.

Each tomato flower is no bigger than a penny. Here's a close-up drawing of a flower.

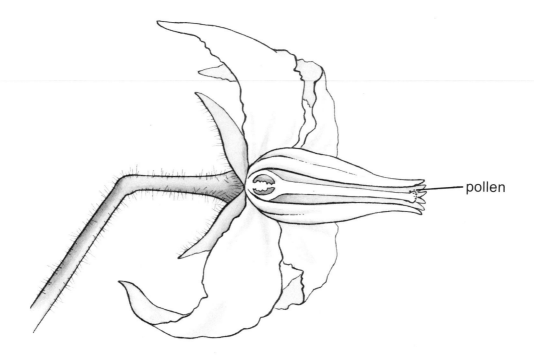

pollen

The flower has been cut in half so that you can see inside it.

Some of the pollen inside the flower falls onto a short stalk in the middle of the flower. When this happens, a tiny tomato may start to grow.

A tomato starts to grow.

Four days after the flower has opened, its petals begin to droop. A tiny tomato starts to grow. The old petals hang down from the tomato.

Look at the big photograph. The tomato is two weeks old. It is the size of a cherry. The petals are dry and soon they will fall to the ground.

The hairy green spikes are the sepals that covered the flower bud. They will stay joined to the tomato.

The tomato ripens.

The tomato grows bigger. It needs lots of water to make it swell. After four weeks it is full-sized, but still hard and green.

Several tomatoes grow on the same branch.

The tomato begins to change color. First it turns orange, then red. Look at the big photograph. Which do you think is the ripest tomato? Which do you think is the least ripe?

The tomato is ready to eat.

Now the tomato is ripe. It is bright red and round and juicy. Look at the small photograph. The little green tomatoes haven't had time to ripen.

You can pick a ripe tomato. It tastes almost sweet. You can keep one of the tomato seeds and plant it.

What do you think will happen then?

Do you remember how a tomato plant grows?
See if you can tell the story in your own words.
You can use these pictures to help you.

3

6

Put an unripe tomato on a sunny windowsill.
How many days does it take to turn red?

Index

This index will help you to find some of the important words in this book.